AF228468

CHIPMUNKS

by Martha London

Cody Koala

An Imprint of Pop!
popbooksonline.com

abdobooks.com

Published by Pop!, a division of ABDO, PO Box 398166, Minneapolis, Minnesota 55439. Copyright © 2021 by POP, LLC. International copyrights reserved in all countries. No part of this book may be reproduced in any form without written permission from the publisher. Pop!™ is a trademark and logo of POP, LLC.

Printed in the United States of America, North Mankato, Minnesota

082020
012021

THIS BOOK CONTAINS
RECYCLED MATERIALS

Cover Photo: Shutterstock Images
Interior Photos: Shutterstock Images, 1, 5 (top), 5 (bottom right), 9, 11, 13 (bottom left), 13 (bottom right), 14–15, 21; iStockphoto, 5 (bottom left), 7; Steve Maslowski/Science Source, 13 (top); Leonard Lee Rue III/Science Source, 16; Tom McHugh/Science Source, 19

Editors: Christine Ha and Brienna Rossiter
Series Designer: Sophie Geister-Jones

Library of Congress Control Number: 2019955007
Publisher's Cataloging-in-Publication Data
Names: London, Martha, author.
Title: Chipmunks / by Martha London
Description: Minneapolis, Minnesota : POP!, 2021 | Series: Underground animals | Includes online resources and index.
Identifiers: ISBN 9781532167607 (lib. bdg.) | ISBN 9781532168703 (ebook)
Subjects: LCSH: Chipmunks--Juvenile literature. | Rodents--Juvenile literature. | Rodents--Behavior--Juvenile literature. | Burrowing animals—Juvenile literature. | Underground areas--Juvenile literature.
Classification: DDC 599.36--dc23

Hello! My name is

Cody Koala

Pop open this book and you'll find QR codes like this one, loaded with information, so you can learn even more!

Scan this code* and others like it while you read, or visit the website below to make this book pop.

popbooksonline.com/chipmunks

*Scanning QR codes requires a web-enabled smart device with a QR code reader app and a camera.

Table of Contents

Chapter 1

Running for Cover

Chipmunks are small **mammals**. They dig **burrows**. They use the burrows to store food and stay safe. Most chipmunks live in North America. Some live in Asia.

Watch a video here!

Chipmunks often make their homes in the forest. They leave their burrows to look for food. They watch for **predators**. If chipmunks sense danger, they hide. They may run back to their burrows. Or they may hide under rocks or logs.

Chapter 2

Chubby Cheeks

Chipmunks have brown fur. They have dark stripes on their backs. Chipmunks also have long, fluffy tails. Sharp claws help them dig.

Learn more here!

Chipmunks gather food and bring it back to their **burrows**. Chipmunks have pouches in their cheeks. The pouches are stretchy. Chipmunks put seeds and nuts inside their cheeks.

A chipmunk's cheeks can stretch to be three times as big as its head.

ear
eye
mouth
cheek
claws

Chapter 3

Two Burrows

Each chipmunk builds two **burrows**. One burrow is **shallow**. The chipmunk uses this burrow during the day. It runs inside to hide from **predators**.

Learn more here!

A second burrow is deeper underground. This burrow is larger. It has many rooms. One of these rooms

is used to store food. The

chipmunk piles seeds and

nuts inside. It eats this food

during the winter.

Another room is for sleeping. The chipmunk puts leaves and grass in this room. The leaves and grass act as **insulation**. The leaves and grass keep the chipmunk warm during the winter.

Chapter 4

Raising Babies

Chipmunks spend most of their lives alone. Female chipmunks have one or two **litters** per year. Most chipmunks give birth in early summer.

Complete an
activity here!

Chipmunks raise their litter in their **burrows**. The babies stay with their mother for eight weeks. Then, they build their own burrows.

Making Connections

Text-to-Self

Have you ever seen an animal eat food? What kind of animal was it?

Text-to-Text

What books have you read about other mammals? What do those animals have in common with chipmunks? How are they different?

Text-to-World

Chipmunks often live near forests. How would this habitat help them stay safe and find food?

Glossary

burrow – a hole that an animal digs in the ground for shelter.

insulation – materials used to stop the cold from entering.

litter – a group of young animals born to an adult animal at one time.

mammal – a type of animal that has hair or fur and feeds milk to its young.

predator – an animal that hunts other animals for food.

shallow – close to the surface.

Index

Online Resources

popbooksonline.com

Thanks for reading this Cody Koala book!

Scan this code* and others like it in this book, or visit the website below to make this book pop!

popbooksonline.com/chipmunks

*Scanning QR codes requires a web-enabled smart device with a QR code reader app and a camera.